The Lotus Birth of the Universe

A Theory of Everything - From Big Bang, to Life, and Beyond

Chandulal Jivandas Rughani

The Lotus Birth of the Universe
Copyright © 2023 by Chandulal Jivandas Rughani

All rights reserved. No part of this publication may be reproduced, distributed,
or transmitted in any form or by any means, including photocopying, recording,
or other electronic or mechanical methods, without the prior written permission
of the author, except in the case of brief quotations embodied in critical reviews
and certain other non-commercial uses permitted by copyright law.

Tellwell Talent
www.tellwell.ca

ISBN
978-0-2288-7936-7 (Paperback)

''I have deep faith that the principles of
the universe will be both beautiful and simple''

-Albert Einstein

The illustrations are volunteered by my skilled grandsons:
Amar, Shaiden, Kian, Kamran, and Raine

Gratitude

I have been blessed with the many ways of support I received in making this book.

I am very grateful to Dr. Don Lincoln for his course "Theory of Everything: The Quest to Explain All Reality". He mentioned the gaps in the theory and encouraged fresh insights! My inspiration was kept kindled.

Ravi, my son, helped me a lot by raising crucial queries prompting deeper insights. I could not have published the book without you. Thank you, Ravi.

The whole family pitched in to help me in diverse ways. Usha listened to my ideas patiently with interest. Neel consistently encouraged me to publish the book. Monica and Priya were helpful in many ways, especially during my bout with the COVID-19 virus. To top it all off, my grandsons contributed to make the drawings for the book! Great!

I thank the staff at Credit Valley Hospital who helped me survive said virus. The nurses also encouraged my endeavour to write the book!

During the challenging times of solitude I had various support from many:

Every morning Priti Abhani emailed beautiful pictures of Hindu deities that pepped up my day.

On Monday mornings, Hasmukh Devani and Ramesh Majithia gave spiritual talks on Zoom, which kept my spirits up.

Regular trips to the ocean and seeing the panoramic views of the mountains and the trees offset the pandemic blues. I am indebted to Jonn and Sue Gaytmenn for providing this wonderful, cosy condominium. All the residents and workers at the condo are very helpful. Thank you all.

Mr Sasha Jordache helped by solving computer problems with a lot of patience. Thank you.

Abdulmohamed Juma has been great to go to the ocean with to shoot the breeze!

Also, countless others have helped me throughout my venture in last few years:

I thank the Mississauga libraries' staff to avail books and DVDs needed for research.

I have talked about my venture to bridge players; their support is like winning a grand slam!

Avnish and Dipti, Sanjay and Nimisha, Lalit Kanabar, Narendra and Dulari and many others who supported me in different ways, I thank you all.

I am indebted to people who offered to help in publishing this book, my heartfelt gratitude to all.

I am lucky to have met Marilyn Parkin in a timely way, to discuss the book prior to its publication. I am deeply grateful for her encouragement.

I want to convey my sincere appreciation to the readers for encouraging me to express my inspirations.

Lastly, the pandemic provided solitude to concentrate for hours, for which I pay my obeisance to the virus from a distance, wearing a mask.

Table of Contents

Preface

A sperm races to the prized egg competing against hundreds of compatriots. It is a Himalayan climb. Compulsive desire to perpetuate HIS legacy drives HIM. SHE is eagerly awaiting the winner, assured of a strong healthy 'catch' for HER future progeny. From a tiny inconspicuous egg/sperm rendezvous, a fantastic baby is kicking to join the world.

Miraculously YOU are born.

Rolling back some 13.7 billion years, when Time itself was not there, the Universe was borne.

Now the Big question is: "How did the Big Bang give birth to the Universe from nothing?"

Buddha claims that there is no solidity in the Universe, "only vibrations, constantly changing, changing". Recently, science has come to the same conclusion. Quantum mechanics in particle physics has reached sub-atomic particles and is digging further. I have followed subjects such as genetics, neurology and psychology to better understand available information at the molecular level. For example, how do genes communicate and coordinate to produce proteins? How do neurons transmit messages? How do humans think? Then mulling over the collective information for days helped to visualize the Big Birth (BB; also used for Big Bang) connection.

The inspirational insights started a few years back. The insights came as brilliant ideas to explore. Usually they happened early in the mornings as I got up. I tried to capture the beautiful floating words by writing them down before they melted away. Yet, I did lose some. Reading it later on, I was amazed – did I really write this?

At first it was just a game to collect all kinds of ideas. The insights were confidently persuasive. After collecting more than 300 pages, I methodically organized them by coalescing each particular topic from many strewn pages. Then the coordinated picture of the birth of the Universe emerged.

Scientific investigation into the beginning of the universe largely follows an "outside-in" approach. This book however starts from the Big Birth up - from Time Zero with No-Thing (the word is used to emphasize the absence of anything, including Space). Of course the ideas kept improving and expanding to include Singularity, Life and the Re-birth of the Universe.

Many a time, watching the courses on DVD, I was sceptical about its validity but was encouraged that the inside-out angle may possibly help the theory. After all, theories are speculations to be tested. Why not express my stubborn exquisite philosophy? It makes sense to publish the theory with confidence.

I have indicated which scientific idea(s) triggered the topic within [] brackets.

I have also written light humour or stories to ease the rendition by using {} brackets.

Let us auspiciously start our journey to the beginnings of the Universe.

> {Vishnu the Preserver, Brahma the Creator and Shiva the Destroyer are the trinity of Hindu deities of cosmic creations.
>
> Some time before Time itself, the Deities floating on the clouds were bored. Vishnu called Brahma and Shiva to do

something about it. The experts prepared a plan to create the Universe from scratch.

As soon as the plan was hatched a trouble-maker, Naradmuni barged in uninvited as was his want. He had a very good idea to spice up the entertainment:

'' we must have Human-beings to provide a good drama''

The plan was stamped!}

Explanatory Notes

- [] Brackets are used to notify the source(s) that triggered the inspiration of the topic.
  {} Brackets are used to tell a light story.
- If the first letter of a word is capitalized, it is related to the working of the Universe.
  Examples:
    - Energy is the fundamental forceful entity, but energy is for ordinary purpose.
    - Mass denotes a virtual entity but mass is a measure of it.
    - Universe, Time, Space etc are mostly used to honour the special entities.
    - Electro-Magnetic (EM), Strong-Weak (SW), Space-Time.
    - Zero, Infinity, Force, Fields, Galaxy etc.

<u>Glossary of Terms:</u>
- Big Bang (BB) and Big Birth (BB) are the same.
- Sphere signifies a growing entity, Globe as a fixed entity (Universe, Galaxies etc.) and Bubble as manifested in the Universe (Atoms, particles etc.). Rarely bubble (with a b) is used as a filling entity (the whole Universe is made of moving bubbles of vibrations).
- Entity is a functional unit made by the patterns of a group of bubbles.
- The word "No-Thing" is used to emphasize emptiness, defying the habitual meaning of Space.
- Universe is the hardware from BB to Mass, and Nature is the software executing the functions.

Overview

This artificial methodology is used to visualize how the Energy, step by step, makes the crucial ingredients to create the Universe. It also explains how the ingredients have to balance themselves with imaginary counterparts. This satisfies the action-reaction principle and compulsive duality to create something from nothing. The artificial slicing of the Universe commences across the middle of the Sphere (The actual creation is described in the following chapters. Also, please refer to the Explanatory Notes section on the previous page for further clarification on the distinction between Sphere, Globe and Bubble).

The Big Birth starts by the Energy making Space (3-Dimensions) and Time ingredients. The upper semi-sphere (Slice 1) shows the ingredients with a plate at the cut of the slice.

The lower half, not shown, is a "ghost" of the upper one, reflecting Energy, Space (3D) and Time. Thus the "ghost" semi-sphere balances the Universe with negative (imaginary) ingredients.

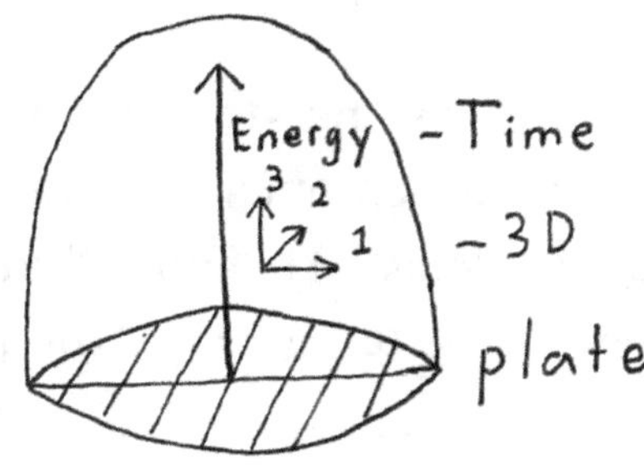

FIG 1: SLICE 1 –
SEMI-SPHERE
Space-Time

The plate: The Energy on the 1st and 2nd Dimensions forms a rotating plate of Electro-Magnetic (EM) force. By splitting off a chunk of Energy, the EM force bends the plate to create the bona-fide Global Universe, which exists independently. The Momentum to bend is the crucial ingredient to form the Global Universe.

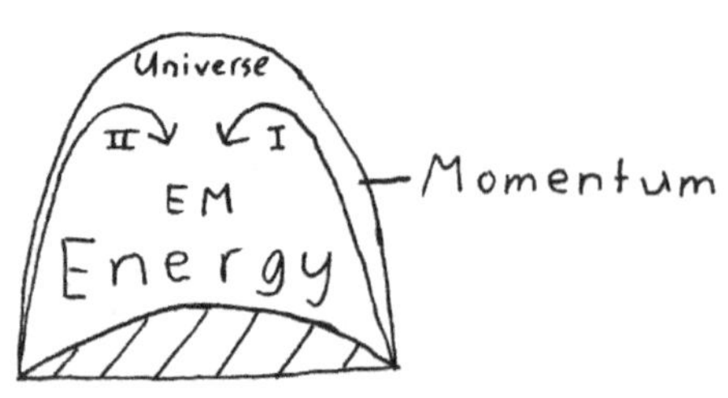

FIG 2: SLICE 2 –
QUARTER-GLOBE
Momentum

The 2nd slice of the quarter-globe shows the Momentum spraying the EM Energy as a hot lava fountain (quadrants I and II) inside the Global Universe.

The mirror-images of the Momentum and the fountains balance the front and back sides of the quarter-globe (quadrants III and IV, not shown).

FIG 3: SLICE 3
Mass

The 3rd slice simply shows Mass in quadrant l and Anti-Mass is in quadrant ll. Mass is like a hill and Anti-Mass is its balancing ditch.

Theoretically the Universal Globe is the first Bubble, itself filled up with Bubbles that perform functions.

It is interesting to note that the Overview follows the steps taken from the Birth of the Universe. It shows the crucial milestones – Space-Time, Momentum of EM force (making Globe) and Mass. Also note that the first quadrant justifies the geometrical 3-Dimension, as explained in Chapters 2 and 3.

The theory posits that the Energy shapes everything in the Universe by vibrations. The momentum and the compactness of the the vibrations (mass) define each 'virtual' Mass. The manipulating of the EM force to create the Global Universe enabled Mass. The whole process is (virtually) manifested from No-Thing to Mass and beyond.

This, in a nutshell, is the Theory of Everything.

It is an honour to the "No-Thing" concept to start the Theory with Singularity labeled as chapter "zero".

Chapter 0

Singularity

This is a very difficult subject to wrap your head around. It is the beginning from No-Thing. It does not even have space or time to start. It is the centre (zero) point of the future Universe; yet without the centre.

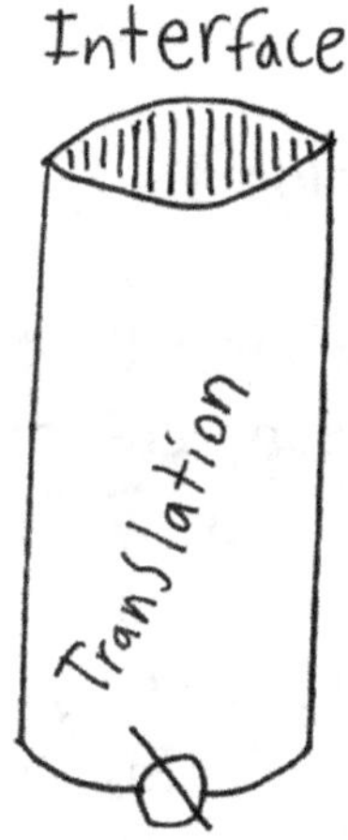

FIG 4: SINGULARITY
MODEL

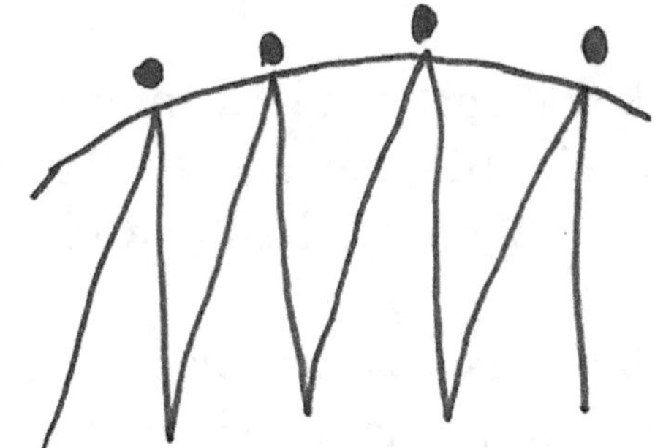

FIG 5: DIGITAL/
ANALOGUE MODEL

The Digital/Analogue model shows two different systems. The Digital information is residing in the Never-Never land shown as Zero (0) in the Singularity model. It contains the Fields of instructions and data in software without the computer. The spiky Digital language signifies erratic behaviour without Space and Time. The Digital language has to be translated to smooth Analogue format (depicted by an arc) upon entering the Big Bang orifice. The Field instructions enable a smooth chronological sequence using freshly made Time.

The Analogue software, upon receiving the coded alphabet [like in genes], executes the Energy Field to bootstrap itself to start making Space-Time. The Energy also spreads along the knitted-net containing nodules of the software Fields. The nodules hold complete sets of the Field's instructions to execute all the functions of the Universe. The resulting data of the executed process is recorded in the nodules. The updated Analogue software is available to Zero in Digital format for posterity. The rhythm of the Universe is smooth since the commanding Fields are available anywhere, anytime.

It is interesting to note that the interface (at the Big Bang orifice) between No-Thing (Singularity) and the Universe can be thought of as a ground for the Quantum Mechanics Theory to tackle the Digital/Analogue conundrum. Perhaps the String Theory may solve it by understanding the code [like mRNA] initiating the vibrations of the Strings.

[Indra's net, Computer software, DNA Technology]

The Zero at Singularity prepares to start the Big Bang. The translation portion is unfathomable. It is like a seed with no concrete contents and no pod; only 'virtual' software.

The Black hole singularity, created at the star's death (supernova), gouges nearby matter. This process is the reverse of the Big Birth, where matter is created from No-Thing.

It is interesting to note that there are two ways to reach the BB Interface: by analytic observation of the Universe [Telescope] and by understanding the sub-atomic particles and beyond [Microscope].

It is also very interesting to note that Telescopic observation can go up to the BB Interface, looking at the Energy shooting in! However, Microscopic examination cannot negotiate the 2 dimensional Inflation period.

Chapter 1

Big Birth

The whole Universe is made of vibrations. There is no solidity. Everything is in flux; just vibrating. Constantly moving, changing.

[Siddhartha Gautama observed the Nature of Universe when he became Enlightened; the word Buddha simply means the enlightened one]

Think of tumbleweed rolling in a dusty wind on a searing hot day in a desert.

The Tumbleweed is actually knitted by vibrations. So are the wind and dust. So also are the sun and its rays. Well, so are YOU watching the scene.

All vibrating bubbles are dancing to the cosmic rhythm of the Universe.

Vibrations, Vibrations, Vibrations.

Energy drives the vibrations. It is the sole currency of the whole Universe – lock, stock and barrel. Energy made the tumbleweed, the sun and YOU. It also provides the forces to blow the wind and send the sunlight. Energy is the maker of mass and motion. It sustains life as well.

Energy is the sole force of creation as shown by Energy Model.

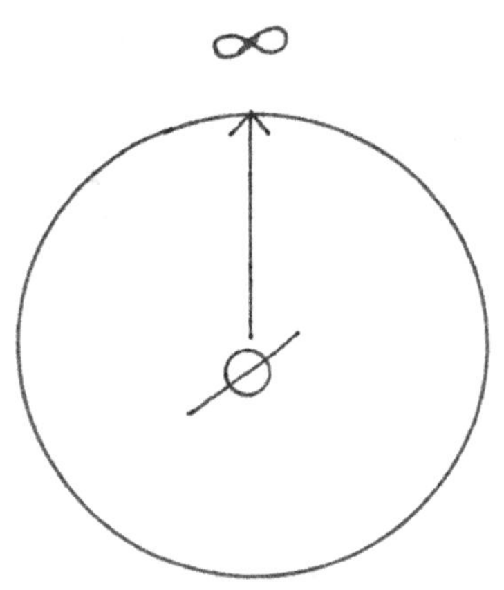

FIG 6: ENERGY MODEL
Energy String

The "string" in the model represents the flow of Energy from Time Zero at Birth to Infinity at the future collapse of the Universe.

(The significance of Zero and Infinity (Appendix A))

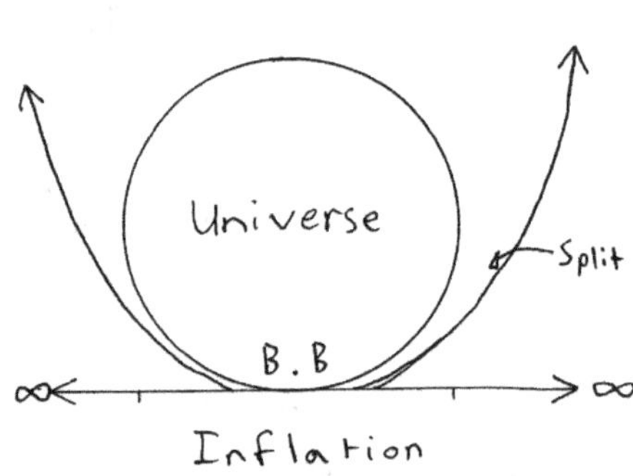

FIG 7: LOTUS MODEL
Split

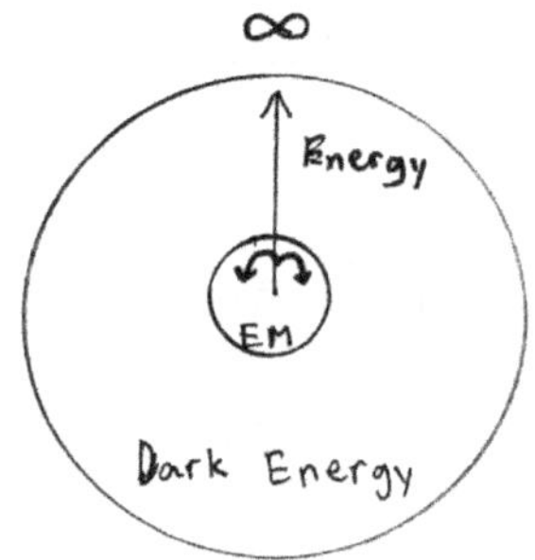

FIG 8: GLOBE MODEL
Energy Flows

{Not yet done to scale, working on it}

The Lotus model describes how the Universe was created. At the Big Bang (BB) the Energy shot out linearly with a strong force at a very high speed. This "Energy-force" refers to the process that created the Universe. This strong Energy-force (1st Dimension) was immediately changed to vibrating Electro-Magnetic (EM) force (2nd Dimension), reaching the end of the Inflation period.

The split of Energy enabled the Strong-Weak (SW) force to shoot back to high speed linear (3rd Dimension) and the EM force to bend by momentum to create the Universal Globe (Chapter 2 describes forces and dimensions in detail). If the Energy had continued at full speed without splitting, it would have kept going to Infinity and the Universe would not have happened. The Globe model shows that the SW Energy force (making Space-Time) continues as Dark Energy.

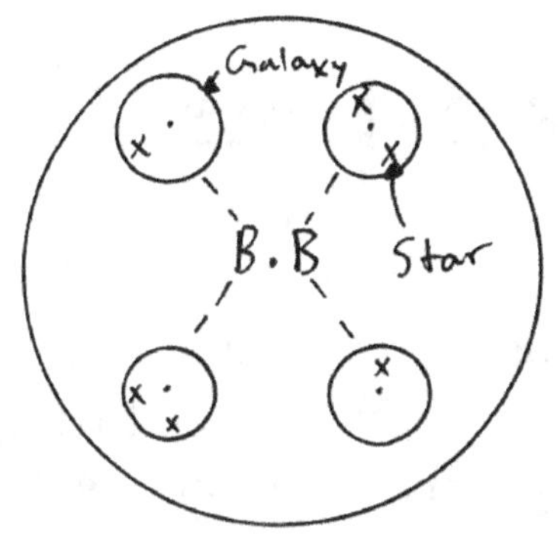

FIG 9: GALAXY MODEL

The molten lava in the Universe formed (future) rotating Galaxies revolving around the BB orifice. There is an umbilical cord connecting the Black Hole at the centre of the (future) Galaxy to the BB orifice. (A Deviation, Gravitational pull: If the Energy were to stop dead from the BB orifice, the Galaxy-cauldrons of molten lava would instantly drop back home to the BB through the umbilical cord).

As the lava cooled, the Stars were born in the rotating branches, which revolved around the umbilical Black Hole of the Galaxy.

[1. Higgs boson creates Mass. 2. Dark Energy has continued to grow from zero at Big Bang to more than 70% of the total available Energy and ongoing]

Chapter 2

Forces and Dimensions

It takes the Energy-force seven steps at 90° dimensions from the Big Bang to make virtual Mass. For simplicity, the flow of the Energy-force is explained as a string to show how it generates vibrations. It is imperative that every step has a balancing reaction to justify the duality.

The translation of the instruction-Fields from digital to analogue at the singularity-interface starts the point-Zero Dimension.

The analogue Energy-Field bootstraps to shoot out Energy with instruction-fields - 1st Dimension. The 1st D Energy-force is a straight linear string at a very high speed, breaking the symmetry. The balancing counter-force has to slow the speed to create the Universe.

The balancing force vibrates the string up and down - 2nd Dimension. The up-down vibrating string itself balances the electrical wave by the magnetic wave, as in a tango dance.

The waves increase in amplitude and frequency (the string can also be thought of as a coiling plate). At the end of the inflation time, three actions happen sequentially:
1. The Energy splits, separating into Dark (SW force) and Light (EM force) Energies to create the Universe.
2. After the split, the SW force continues as a high speed straight linear string -3rd Dimension. The balancing tag of the weak force enabled the Sphere to continue as Dark Energy above Light Energy.
3. The EM component of the split provides momentum to bend up to the 3rd dimension, creating the Universal Globe.

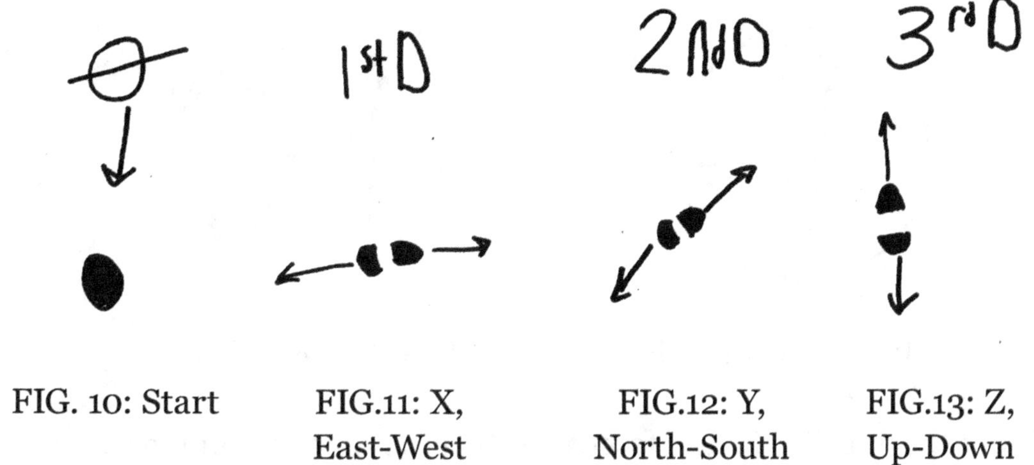

FIG. 10: Start   FIG.11: X, East-West   FIG.12: Y, North-South   FIG.13: Z, Up-Down

The three Dimensions are related to Electro-Magnetic (EM) and Strong-Weak (SW) forces.

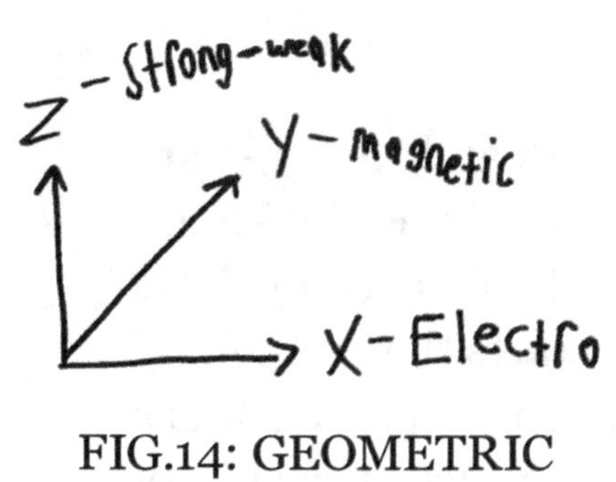

FIG.14: GEOMETRIC 3D MODEL

1st Dimension – Electric
2nd Dimension - Magnetic
3rd Dimension – Strong-Weak

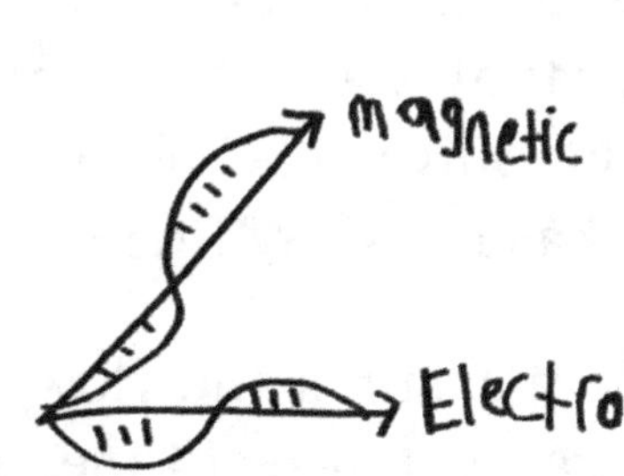

FIG.15: ELECTRO-MAGNETIC FORCE

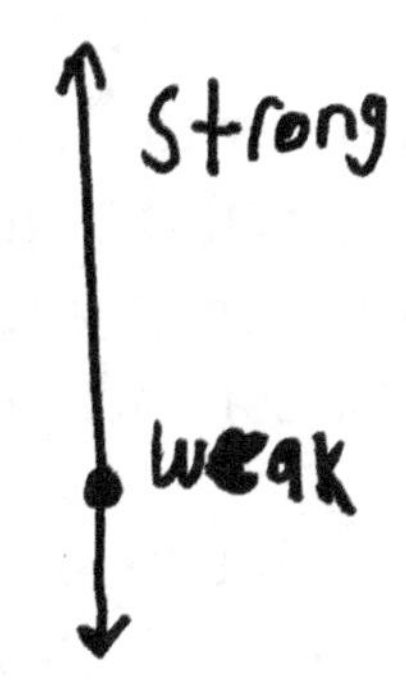

FIG 16: STRONG-WEAK FORCE

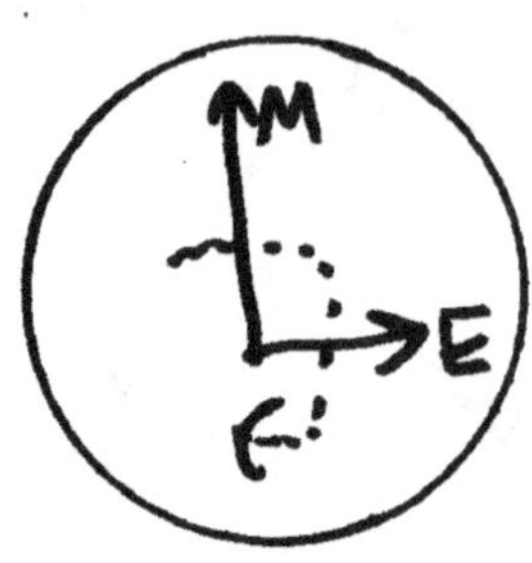

FIG 17: EM PLATE

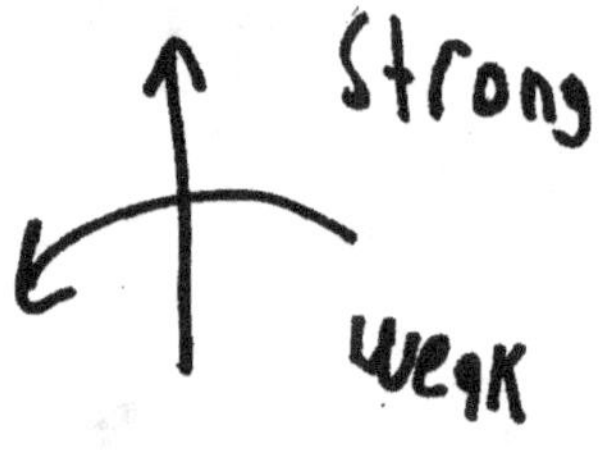

FIG 18: SW MODEL

More specifically, the EM force travels at the speed of light in a linear track - 5th Dimension. The track itself balances the Light and Dark sides to create the bona-fide Universal Globe. The EM plate caps the Globe with the 2D vibrating instruction-fields carried by the shooting EM Energy. The SW force continues the process of Space-Time on the 3rd dimension, going up.

It is interesting to observe that steps 1-3 and steps 5-7 are similarly correlated to both Pythagoras' and Einstein's equations (The 4th step, measures the Time taken by these steps [no pun intended]. Also step 0 starts the ball rolling by breaking the symmetry). Pythagoras' equation takes care of three steps, the dimensions of Space. The split of Energy prepares for Einstein's equation, as explained in the next chapter.

It is interesting to note that the 2D EM wave is carrying the instructions (vibrations) to enable virtual Mass. Also note that the Strong initial Energy flows through the Universal Globe to help create Atoms (entities) and continues beyond as Dark SW Energy.

The 4th Dimension, Time, simply measures the flow of Energy through Space. The "speed of light" is a reference of time, like a scale on a map. Much like a clock, it is used by the Universe to synchronize and coordinate its functions.

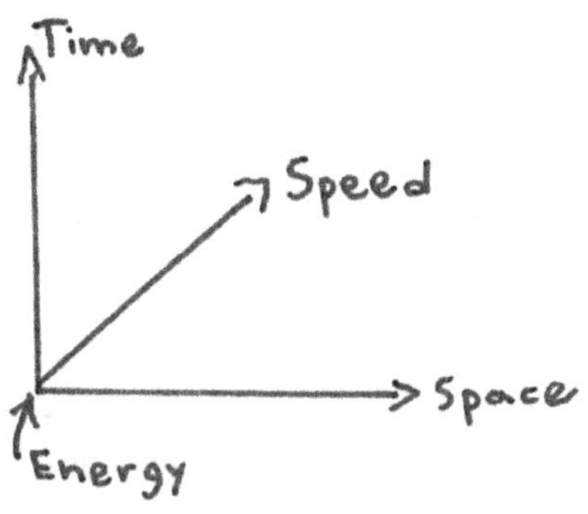

FIG.19: SPACE-
TIME MODEL

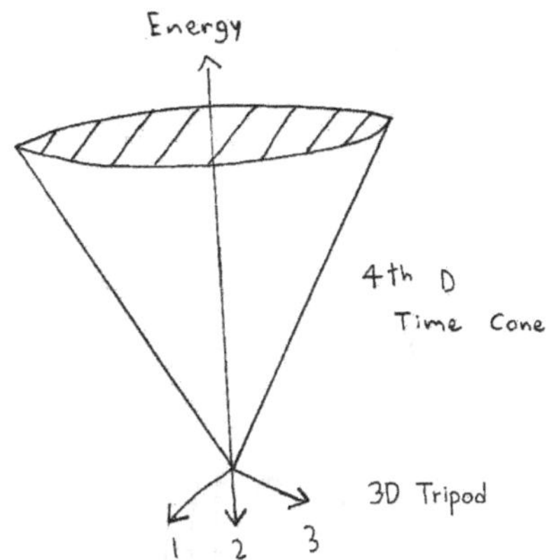

FIG.20: PICTORIAL

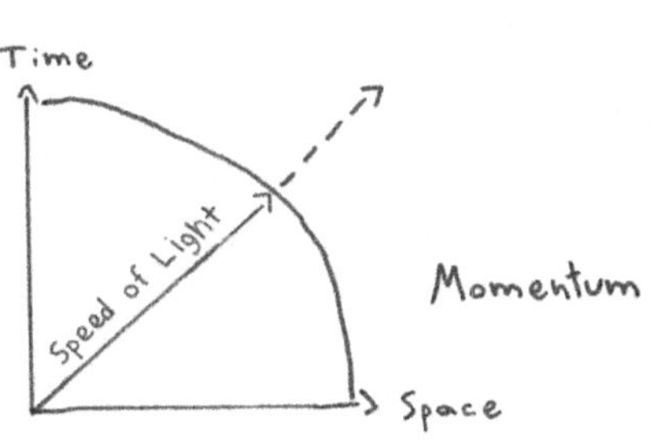

FIG. 21: MOMENTUM
MODEL

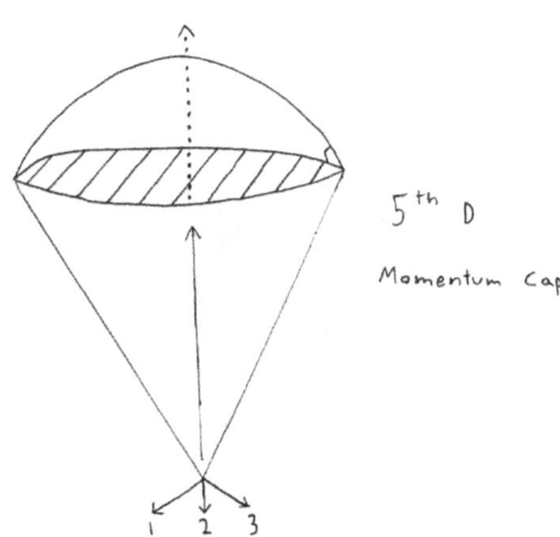

FIG. 22: PICTORIAL

The spraying down of the EM-force in the 3rd Dimension is balanced by the Spin of the Globe - the 6th Dimension. The shooting straight down linear string itself is balanced by the tethering induced because of the Global circularity. This creates a Universe of hot lava.

In time, the virtual Mass is balanced by the cooling of lava – the 7th Dimension. The pressing of the tethering is induced by the Dark Energy Sphere. At last, the 2D EM force enabled the Mass vocabulary to create Fields to execute Atoms, etc.

The next chapter describes the details of Mass.

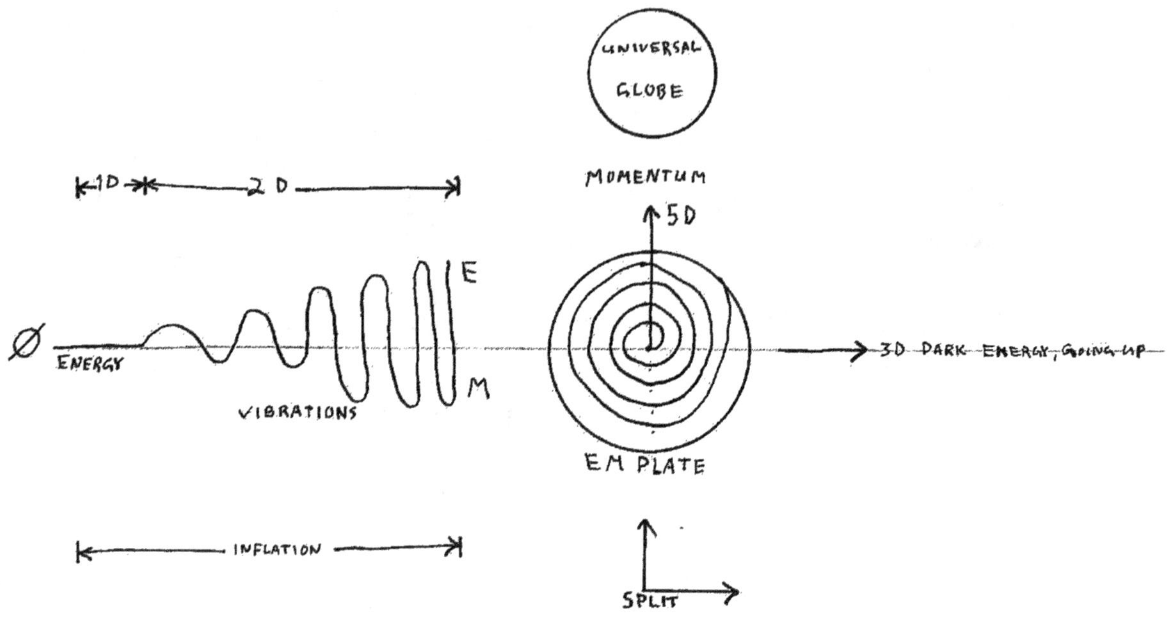

FIG.23: UNIVERSE MODEL

Chapter 3

Mass

The split into EM and SW forces enabled the Momentum to create the Universe with the complete Space of three dimensions. The Momentum used its EM force, at the speed of light, to define the boundary between Light and Dark areas.

The best way to describe Mass is to superimpose on the Momentum model from the last chapter. This is like combining Einstein's and Pythagoras' equations.

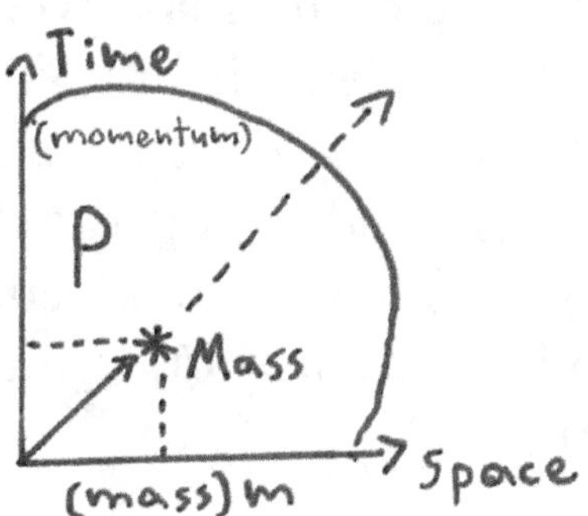

FIG.24: MASS MODEL

Einstein's Pythagorean equation, $E^2 = (mc^2)^2 + (pc)^2$ describes the model perfectly. Mass shares the Energy (E) between its momentum (p) and its mass (m). In other words, the faster the spinning momentum (squeezing energy), the thicker the mass when slowing down the speed of light.

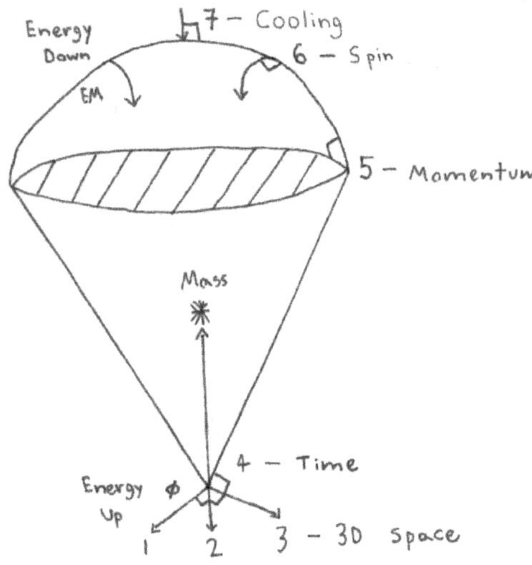

FIG.25: PICTORIAL

Also the Time utilized by the Mass in the Universe is explained by the constant speed of light (c). It is interesting to note that m is multiplied by c² in the equation perhaps justifying positive and negative masses, while the p is multiplied by c means no negative momentum in "ghost" area (Balanced Mass Model).

The Mass Model shown above is simply a 2D representation of Mass.

The creation of Mass from No-Thing has to balance its mass, momentum and spin as shown in the composite Balanced Mass model.

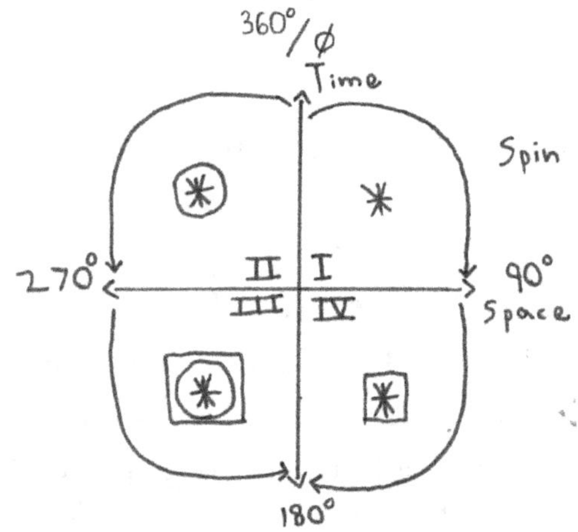

FIG 26: BALANCED
MASS MODEL

Mass (starred*) in 1st quadrant has positive spin and mass while Anti-Mass (circled) in the 2nd quadrant has negative spin and mass. The 3rd and 4th quadrants are simply "ghosts" balancing Mass and Anti-Mass quadrants (squared).

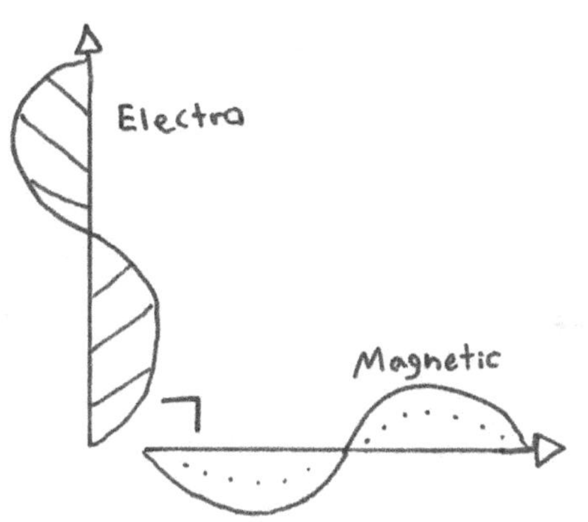

FIG.27: E & M WAVES

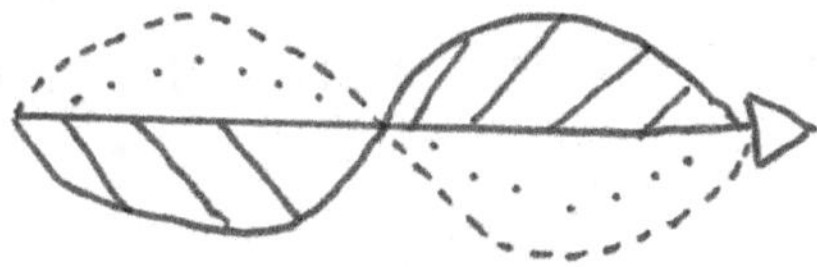

FIG.28: EM WAVE

The EM wave is a composite wave made of it's respective components -Electro and Magnetic. Like an arrow, it is perfectly symmetrical in design. The EM force is the keeper of the instruction Fields.

It is interesting to note that the EM force (2D) wave itself justifies the balancing phenomena, like an arrow. The wave follows the spin from the top of the Balanced Mass Model clockwise and anti-clockwise. The crest of the wave (90 deg.) is Mass, while the trough of the wave (240 deg.) is anti-Mass.

FIG.29: EM WAVE
MODEL

{The creation of Mass enables our existence and our environment. On July 4th 2012, scientists discovered the Higgs Boson -a very elusive particle that enables Mass. No wonder the Higgs Boson was dubbed "God's particle".}

Chapter 4

Fields

(Roughly speaking, the Universe is the rock bottom hardware from BB to Mass, while Nature is the software providing the efficient functioning of the Fields).

Like the software in a computer, all the Fields are sophisticated programs to take care of executing all aspects of making the Universe itself and running it. The software also collects all the data of the events happening in the Universe. Initially all the programs in the software lie dormant in Zero (chapter 0). Upon Big Bang, the translated Energy Field bootstraps itself to start shooting Energy. The Energy also carries the software in the nodules of a net. Thus, the Fields are available anywhere at any time. Any events happening anywhere in the Universe are instantly recorded and available everywhere in these nodules.

Communication is executed in wave-shaped words by the involved particles, which themselves are wave-shaped.

As an example, Higgs Field gives an allocated mass to sub-atomic entities through the Higgs boson particle. Both Higgs Field and Higgs boson particle (communicating carrier) are wave shaped. So are the nucleus and electrons of the atoms. The whole Universe is comprised of vibrating bubbles, amicably functioning together. The buildup of bubbles create heavy-weight rotating entities (Planets) along charted geographical tracks.

[Indra's net, Computer software, Particle Physics]

Chapter 5

Life

The story of the birth of a baby at the beginning posed a question: how did the Universe start?

Now it begs another question: how did human life start?

The theory of the Big Bang implies that the Fields enabled creating the Universe. In a similar way, Nature may have Life Fields to give biological birth. By extension, the Fields may also provide necessary functions to survive in the habitat, which were also made by associated Fields. The Life Fields, gradually changing species by evolution, created human beings. Endowed with intelligence, humans understand the rudimentary workings of Nature itself. However, overall our knowledge of Nature is still limited.

Currently two important issues need to be addressed urgently for the human species to survive by fully complying with Nature:

1. Carbon is a versatile entity in the Universe. As a strong element, it provides solid structures such as bones, necessary for life. Trees take in carbon dioxide in exchange for oxygen from air. It also provides food for energy! It is win-win situation. Could the depletion of carbon be catastrophic by tempering with fossil fuels, deep down in the Earth for eons? Could the fundamental balance of breathing fresh air be compromised? Understanding Nature's way to manage the chain of carbon's function may enable us to find amicable solutions.

{A whistle of the first commercial train in England in the nineteenth century heralded the Industrial era. Lots of progress has helped human endeavors. Yet the exponential industrial growth over a century has unbalanced Nature's environment}

2. Viruses are one of the earliest biological entities with very limited features but a very spunky survival instinct and a quick response to adverse conditions. Although viruses cause a host of infections in humans, they also provide for potential medical benefits. For example, clinicians now use oncolytic viruses to target and kill certain cancerous cells. Yet this war between humans and viruses, if understood Nature's way, may provide equitable solutions.

{Life is a precious gift – a jewel. We are endowed with such a brilliant mind to make our existence a joy! Arts, language, mathematics are examples of our skills. Culture, family, friends, etc. are our supports. Music, dance, games, etc. entertain us. Trees, water, environment, etc. are our livelihoods. Please take care.}

It is interesting to note that there must be some reason for Nature to create humans to do something.

It seems that a human being is given Time to _______________________.

What do you think? What is the meaning of *your* life?

Chapter 6

Rebirth

When the Energy reaches its predetermined limit, the Sphere reaches Infinity. Time has come to an end. It is time to collapse to Zero.

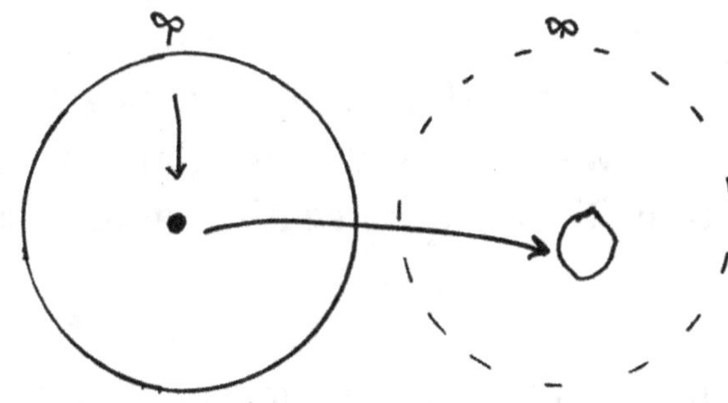

FIG.30: GLOBAL MODEL
Rebirth

The left side globe is at steady state prior to collapse. The right side is the start of the rebirth at Zero.

The Big Birth commences. The protocol is well established. The software is available from the previous incarnation. From the BB slit, the Energy starts the ball rolling. The rest is history.

[It is known in Particle Physics that the Universe is a huge smooth slab (flat, zero curved) with very negligible contents. However there is a very small slit discerned in one spot, maybe for BB?]

Appendix A
Zero and Infinity

The start of the Big Birth at zero Time is difficult to comprehend. Dualities of up and down, left and right etc are nonsensical to a starting point with no dimensions and no concept of time or motion. Still zero is the starting point from nothing. To top it all off, the information (in squiggly digital language) sitting in zero has to pass it on in smooth analogue vernacular. Is the zero then valued as a positive-negative demarcation? Of course not! The best way to describe singularity zero is to tie it with negative Time. Call it Ghost since negative Time does not exist.

Zero helps to balance all things as they happen to keep a clear account of all transactions.

Zero is also the interface to transmit information both ways.

The shooting Energy starts at Time Zero.

Infinity is reached when the Energy ceases. All the motions will stop - a steady-state. Sudden collapse of the Universe occurs. It is time to go back home, to Zero.

Rebirth starts.

Zero and Infinity are themselves end blocks of the biggest Bubble. It is interesting to know that both end blocks are inaccessible, yet give existence to all the entities of the Universe.

Homage to Siddhartha Gautama

In the twenty-first century, Siddhartha Gautama is universally recognized as one of the greatest multi-skilled geniuses of all time. Many professionals in fields like science and psychology have used his teachings. Scientists discovered that there is no solidity in the Universe - which Siddhartha knew twenty five centuries ago. Psychologists have used mindfulness meditation to help patients manage their illness. He gave practical advice in many facets of life; such as politics, warfare, education, family life, business etc.

How did Siddhartha acquire his knowledge twenty five centuries ago? At a young age Siddhartha felt that no one should suffer from illness, aging and death; 'suffer' being the operative word. Meditating very deeply, he developed insights into the laws of Nature. Thus, Siddhartha Gautama is called the Buddha, the Enlightened One.

Ironically, he learned that suffering is a condition of existence. He taught to simply accept this reality. His teachings are based on 'Four Noble Truths'. These Truths empower anyone interested in reality. The Fourth Noble Truths guide one to lead a Noble life. Noble means it is leading to Truth.

The followers of the Buddha are monks and nuns as well as householders. Monks and nuns endeavour full-time to achieve enlightenment. For centuries they have preserved the teachings. They help householders in understanding Buddha's teachings. Householders, in turn, traditionally support them. All follow the well documented teachings of the Buddha. Anyone can improve their life by practicing the teaching.

In describing his vision of the Universe the Buddha said "There is no solidity, only vibrations. Everything is in flux, changing, changing".

It is important to recognize that the Buddha saw the Universal Truth. With compassion he taught the Noble Truths as a science to be passed on by his followers.

In A Nutshell...

In the beginning, the Energy-Field from No-Thing (0*) bootstraps Energy to shoot at a high speed in a very short Inflation period (Space-Time: 1, 2, 3, 4). Then the splitting of Energy enables the rotating Electro-Magnetic (EM) force to bend, creating a bona fide Universal Globe (Momentum: 5). The EM waves spread down the Globe like a fountain of lava (Spin: 6). Finally, the EM force enables the vocabulary of Mass by Dark Energy (Cooling: 7).

*Pictorial model, in chapter 3, describes 0 to 7 steps to create Mass.

The molten lava cools, jelling into Galaxies, the first entities to form after the Universal Globe. The upcoming Strong force nourishes the Galaxies through an umbilical cord. The SW force continues to make Space-Time beyond the Universal Globe, known as Dark Energy.

The Stars are the condensed energy entities serving surrounding lighter entities.

The rotating Planets revolve around their mentor Star's etched tracks in the cosmos.

The moons follow similar dynamics of their Planets.

The whole Universe is packed with all sorts of vibrating Bubbles (large and small, simple and complex; entwined with each other to perform their complex functions) constantly moving and forever changing. Even the so called empty space is filled with bubbles to support the entities. The Universe hums like the rhapsody of an orchestra on an ever changing, moving platform.

The Lotus model representation, in chapter 1, opens wide to easily describe the functions of the entities of the Universe (it may help to visualize the moving entities, packed bubbles, as vibrating waves communicating to perform the functions. Remember that the Universe is virtual, a mirage):

The boundary, or the cap, of the Universe contains the Fields to enable:
1) Creating the Universe itself
2) Communicating Fields with all entities
3) Separating the boundary of Dark Energy
4) Giving 'speed of light' standard - a scale of Space measure (as described below)

The virtual Mass is developed from the 2D EM force to enable a 3D entity, using momentum and mass (Mass model, Chapter 3). The Lotus model clearly defines the Universe from Big Bang to Mass. The Mass is a crucial stage of development to implement complex structures to make Atoms (It is interesting to note that Mass explains the string theory of Quantum Mechanics and by corollary the Atom Bubble explains the Standard Model).

Time is an illusion:
The waves of the shooting Energy are actually vibrating up and down in 2 dimensions. The human mind assumes the waves to be moving in the 3rd dimension. A surfer on the wave thinks he is on a wave moving sideways. Or the wave convinces the mind that the wave itself moved sideways. The way for the surfer to understand reality is to mentally hold on to the top of the surf, moment to moment. The present Time is also an illusion. It is only arrested for a moment and immediately fools the mind to think of future or past, (front or rear side of the wave). The 4th Dimension is the Time and the 3rd Dimension is the allocated Space of the Globe. (Momentum Cap (Chapter 2) is the 5th Dimension).

The arrow of Time is a misnomer. The only arrow is Energy's circulation of the bubble.

> In relative reality, Time is needed to measure motion. If the Mass, for example, consumed its allocated Energy to motion only, it will lose its mass. If it allocates it all to mass, it stays as a stationery lump. If the Energy force pushes an entity, the motion measures the covered space related to the scale. Thus the speed is a portion of the scale – Time taken.

The instruction Fields create the Universe and also furbish the entities to function. It is interesting to speculate that the Conscious Fields are there to provide biological entities and to enable their evolution. The Mind Field covers a huge function to control the body made from the dust of the stars!

This chapter succinctly describes a mammoth undertaking of over 12 years. It provides insight into the dynamics of the Universe in motion. These inspirational insights get deeper as time passes. All the chapters as written defined the current understanding at that given time. I think that this chapter contains the gist of how the Universe was created.

I had a dream....
A dream that
    our lives must have a purpose
A dream that
    we need to know Nature
And inspiration to shatter our ignorance
    May this be a present to the world

Tidbits

Reflections

Now our journey brings us to take stock of our accomplishment. In the beginning, the ideas were loosely floating around. The meditative thinking on varied subjects jelled to concentrate on the meaning of life. Who am I?

The inspirational thinking pointed to start with the birth of the Universe. Rest is history.

The experience of going through the virtual scenario of the creation of the Universe is mind boggling.

The elegant Universe has endowed humans a privileged position to share in shaping future progress.

Human life has been endowed with more perks than other species. It is a privilege to co-operate with Nature.

It is intriguing that science has significantly furthered our understanding of the Universe, but less so for Nature. Being human is an opportunity to intellectually justify our existence. To be in accord with Nature is fundamental to survive. Ignorance leads to catastrophic mistakes. Science needs to develop the knowledge of connecting Conscious Field to mind.

What I learned most of all is that life is a beautiful present. Enjoy it, every moment!

Paradigm Shift

Any earth-shattering idea is bound to upset a majority of the people ensconced in their beliefs. Generations of embedded social behaviour and genetic endowments shape culture. An idea that the earth revolves around the sun was taken as blasphemy. Surely the sun rises in east and sets in the west. Scientific observations using the telescope confirmed this reality. Also the microscope enabled many scientific discoveries in microbial science.

Going forward, the Fields and their connection to the mind will be a very interesting field in science (no pun intended).

Living well is a blessing

With happiness forever

Loving care gives

Joy to all forever

Never hurting any one

Delights forever

Sharing my Expression

I thank you all forever

Expression is, I believe, the purpose of life. Do not hurt anyone, try to care for the needy and purify the mind are the ways of Noble living. So says Siddhartha Gautama.

May all sentient beings be happy!

Nature is Paramount

THRUVE
North Star

www.ingramcontent.com/pod-product-compliance
Lightning Source LLC
Chambersburg PA
CBHW080525030726
47592CB00012B/3476